Simple Solutions.

Minutes a Day-Mastery for a Lifetime!

Level 2

Science

McGraw, Tondy, Toukonen

Bright Ideas Press, LLC
Cleveland, OH

Simple Solutions Level 2
Science

Printed in the United States of America

ISBN-13: 978-1-60873-020-9
ISBN-10: 1-60873-020-4

Cover Design: Dan Mazzola
Editor: Kimberly A. Dambrogio
Illustrator: Christopher Backs

Note to the Student:

We hope that this program will help you understand Science concepts better than ever. For many of you, it will help you to have a positive attitude toward learning these topics.

Using this workbook will give you the opportunity to remember topics you may have learned before. By going over these topics each day, you will gain confidence in Science.

In order for this program to help you, it is extremely important that you do a lesson every day. It is also important that you ask your teacher for help with the items that you don't understand or that you get wrong.

We hope that through Simple Solutions and hard work, you discover how satisfying and how much fun Science can be!

Lesson #1

Weather

Is it sunny today? Is it raining or snowing? Is it windy? Is it hot or cold? What the air outside is like is called **weather**. Weather can change in a few hours or in a few weeks.

1. _____ is what the air outside is like.

   Wind   Snow   Clouds   Weather

2. Name **two** kinds of **weather**.

   _____

   _____

3. It can be hot or _____.

4. Weather can _____ in a few hours or in a few weeks.

A time of the year that has a certain type of weather is called a **season**. There are four seasons. The seasons are **spring**, **summer**, **fall (autumn)**, and **winter**. The seasons happen in this same order every year. Each season has its own weather.

5. A time of year that has a certain type of

   weather is called a _____.

6. There are _____ seasons.

7 – 8.   List the seasons.

_____

_____

_____

_____

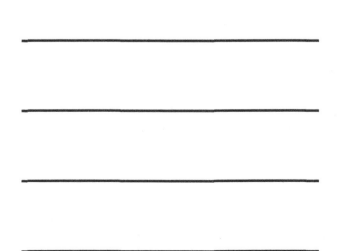

Lesson #2

Tools to Measure Weather

Scientists use tools to measure weather. A **thermometer** measures the **temperature** of the air. The temperature tells us how hot or cold the air is.

Scientists can also measure **wind**. Wind is moving air. It can move in many directions. Scientists use a **weather vane** to measure which way the wind is blowing.

1. Which tool do scientists use to measure temperature?

   _____

2. Moving air is called _____.

3. Which tool do scientists use to measure the direction of the wind?

   thermometer   weather vane

4. How many seasons are there? _____

5. Which thermometer shows the temperature on a warm day?

A)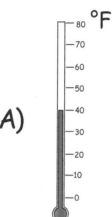

B)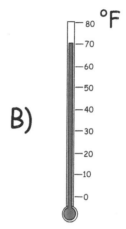

6. Fill in the missing seasons.

__winter_____

_____

__summer_____

_____

7. Name a kind of weather. _____

8. _____ is what the air outside is like.

A) Wind

B) Weather

C) Seasons

Lesson #3

Flood

When a lot of rain falls, it can cause a **flood**. A flood happens when rivers and lakes get too high and overflow their banks. The water covers the land. Houses, animals, and even whole towns can be destroyed by floods.

1. When rivers and lakes overflow, what can happen?

a _____

2. Underline kinds of weather.

windy   temperature   rainy   snowy

3. Which of these tools measures wind direction?

thermometer   weather vane   scale

4. A time of year that has a certain type of weather is called a _____.

5 – 6. List the **four** seasons.

_____

_____

_____

_____

7. Circle the letter of the thermometer that shows the **coldest** temperature.

A) 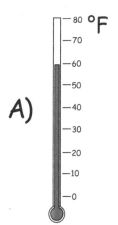   B) 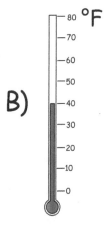   C)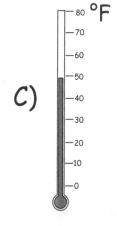

8. Circle the letter of the animals that have hair or fur.

A)    B)    C)    D)

Lesson #4

Drought

A flood happens when there is too much rain. A **drought** happens when it has not rained for a long time. The land gets very dry and hard. Plants may die. Animals have a hard time finding water to drink.

1. A _____ can happen when there is too much rain.

2. A _____ happens when it has not rained for a long time. The land gets very dry and hard.

3. Match each adult animal to its baby.

dog   fawn

bird   puppy

horse   chick

deer   colt

4. Fill in the missing seasons.

_____

<u>spring</u>_____

_____

<u>fall (autumn)</u>_____

5. Which **two** animals make their home in the forest?

squirrel   elephant   deer   penguin

6. What word tells what the air outside is like?

rainy   wind   weather   thermometer

7. Which of these tools measures temperature?

thermometer   weather vane   scale

8. Which animals have feathers?

robin   fox   owl   pig   chicken

Lesson #5

Think Like a Scientist (Part 1)

- Use your five senses to **observe** the world around you.

- You may have a **question** about what you are observing.

- Make a **guess** about a possible answer to your question.

1. What do you use to observe the world around you?

    your _____

2. You may have a _____ about what you are observing.

3. Make a _____ about a possible answer to your question.

4. Which of these tools measures wind direction?

    thermometer   weather vane   scale

5. Match each word with its definition.

    ____ wind   A) no rain for long periods

    ____ drought   B) what the outside air is like

    ____ flood   C) moving air

    ____ weather   D) when rivers and lakes
                          overflow onto the land

6. Which **two** animals make their home in water?

    whale   turkey   dolphin   bat

7. Underline kinds of **weather**.

    cloudy   weather vane   snowy   sunny

8. Match each adult animal to its baby.

    cat   piglet

    pig   cub

    bear   kitten

Lesson #6

Think Like a Scientist (Part 2)

- To find the answer to your question, do an **experiment**. You can see if your guess was correct.

- The last thing to do is to **share** what you learned from your experiment. You can write or draw pictures about it.

1. To find an answer to your question, do

   an e_____.

2. When you think like a scientist, the last thing to do is to s_____ what you learned.

3. What does this tool measure?

   weight   wind direction   temperature

4. You use your s_____ to observe the world around you.

5 – 6. Color the leaves that have a jagged or pointy edge.

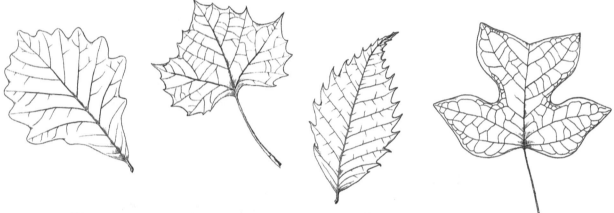

7. Write **T** if the statement is true or **F** if it is false.

_____ An ocean is a large body of water.

_____ Plants and animals are living.

8. Which type of weather is shown?

   A) rainy
   B) snowy
   C) sun

Lesson #7

Living Things

Living things need food, water, and air. Living things grow and change. They can make new living things that are like themselves. Plants and animals are living things.

1. All living things have **three** needs. List them.

    _____

    _____

    _____

2. Write **T** if the statement is true or **F** if it is false.

    _____ Living things make new living things that are like themselves.

3. A weather vane measures _____.

    wind direction   temperature   wind speed

4. Color the leaves that have a **smooth** edge.

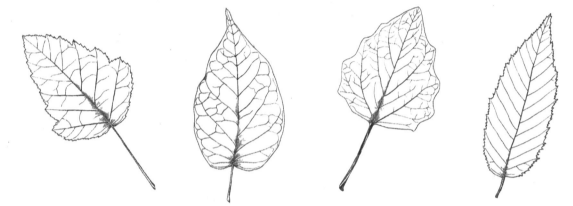

5. Circle the part of the plant where the seeds and fruit are made.

stem   leaf   flower

6. What is it called when the land gets dry and hard because it has not rained for a very long time?

flood   tornado   drought   blizzard

7. Which bird would use its beak to sip nectar?

A)    B)    C)

8. Which season follows winter? _____

Lesson #8

Nonliving Things

Nonliving **things** do not need food, water, and air. Nonliving things cannot make things that are like themselves. Rocks, air, water, and sand are examples of nonliving things.

1. Give 3 examples of **nonliving** things.

_____

_____

_____

2. What is it called when the rivers and lakes overflow onto the land?

    flood   tornado   drought   blizzard

3. Which season follows summer?

_____

4 – 5. Color the leaves that have jagged edges **green**.
Color the leaves that have smooth edges **yellow**.

6. How are the things in the picture below the same?

A) They both are living.

B) They both need water.

C) They both need food.

D) They both are nonliving.

7. All living things need _____.

A) food, teeth, and water

B) air, food, and water

C) food, feet, and water

8. What kind of weather is shown here?

A) thunderstorm

B) blizzard

C) tornado

Lesson #9

Review

In order to act like a scientist:

1. Observe and ask questions.
2. Make a guess.
3. Do an experiment.
4. Share what you learned.

1. When acting like a scientist, what is the next step you should do after you **make a guess**?

   A) Share what you learned.
   B) Observe and ask questions.
   C) Do an experiment.

2. When acting like a scientist, what is the next step you should do after you **do an experiment**?

   A) Observe and ask questions.
   B) Share what you learned.
   C) Make a guess.

3. Which season follows fall (autumn)?

   spring   summer   winter

4. A weather vane measures _____.

    wind direction   temperature   wind speed

5. All living things have **three** needs. Food and water are two of them. What is the third need?

    _____

6. Write T if the statement is true or F if it is false.

    _____ A grasshopper is an example of a nonliving thing.

7. _____ is what the air outside is like.

    Wind   Weather   Seasons

8. How are the things in the pictures alike?

    A) They are both nonliving.
    B) They both have beaks.
    C) They both are living.
    D) They both have a stem.

Lesson #10

1. In science, we use tools to help us observe, measure, or study objects. One of these tools is used to magnify, or make something look larger. It is called a **hand lens**.

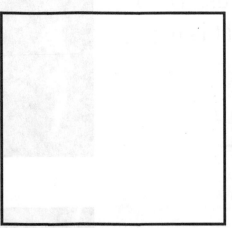 Draw your own picture of a hand lens in this box.

2. What happens during spring?

   A) Trees have no leaves.

   B) Some leaves change color.

   C) Trees grow new leaves.

3. Match each animal with its shelter.

   bee   nest

   bear   hive

   bird   cave

4. Name **two** examples of **weather**.

_____

_____

5. Write T if the statement is true or F if it is false.

_____ A thermometer tells how hot or cold the air is.

6. Which of these are **nonliving** things?

7. What is it called when the land gets dry and hard because it has not rained for a very long time?

flood   thunderstorm   drought   blizzard

8. All living things need air, food,

and _____.

Lesson #11

What Animals Need

All animals need **food**, **water**, and **air** to stay alive. The bigger the animal, the more food and water it needs. Animals also need shelter. **Shelter** is a safe place for animals to live. Some animals find shelter in a tree or in a hole in the ground. Other animals use caves to keep them safe.

1. Animals need **four** things to stay alive. They need food, water, and air. What else do they need?

   _____

2. Name a kind of shelter for an animal.

   _____

3. Spring and summer are seasons.
   List the other **two** seasons.

   _____

   _____

4 – 5.   Match each word with its definition.

_____ flood   A) no rain for long periods

_____ wind   B) what the outside air is like

_____ drought   C) moving air

_____ weather   D) when rivers and lakes
                         overflow onto the land

6. What do you call a tool that is used to magnify, or make
    something look larger?

        weather vane   hand lens   thermometer

7. Write **T** if the statement is true or **F** if it is false.

    _____ A shelter is a safe place for an animal to
            live.

    _____ Rain and snow are types of seasons.

8. What does this tool measure?

    A) wind direction
    B) temperature
    C) rainfall

Lesson #12

1. Which type of weather is shown?

    drought   flood   blizzard

2. Name these tools.

    A) _____

    B) _____

3. Why do animals need shelter?

    for safety   for food   for oxygen

4 – 5. Put these steps in the correct order.
        Number from 1 – 4.
        (See Lesson #9 for help.)

_____ Do an experiment.

_____ Observe and ask
         questions.

_____ Share what you
         learned.

_____ Make a guess.

There are three main types of clouds.

- **Cirrus clouds** are the highest clouds in the sky. They look like wisps of hair.

- **Stratus clouds** are the lowest clouds in the sky. They look like a sheet or layer of clouds.

- **Cumulus clouds** look puffy. They can bring strong storms.

6. Which type of cloud looks like wisps of hair?

   cirrus   cumulus   stratus

7. Which type of cloud looks puffy?

   cirrus   cumulus   stratus

8. Which type of cloud looks like a sheet?

   cirrus   cumulus   stratus

Lesson #13

What Plants Need

Plants need **light**, **water**, and **air** to grow. They cannot move around like animals do. They get the things they need right where they grow. Plants

also need **nutrients** (**food**). Nutrients come from the . They help a plant grow.

Different plants need different amounts of water or light. Some plants, like a cactus, do not need much water. Weeping willow trees can take in a lot of water. Some plants, like sunflowers, need a lot of light to grow, but ferns don't need much light.

1. Plants need food, water, and air. They also

   need _____.

2. Nutrients come from
   the _____.

   A) air

   B) water

   C) soil

3. Write **T** if the statement is true or **F** if it is false.

    _____ Plants get the things they need right
             where they grow.

4. Which type of cloud is shown?

    cirrus   cumulus   stratus

5. Which step comes after **observe and ask questions**?

    A) Do an experiment.

    B) Share what you learned.

    C) Make a guess.

6. Which of these is **nonliving**?

          bear   daisy   air

7. Name a kind of shelter for an animal.

    _____

8. All living things need food, air, and _____.

          feet   soil   water

Lesson #14

1. Which type of cloud looks like wisps of hair?

   cirrus   cumulus   stratus

2. Animals have **four** basic needs. Fill in the missing needs.

   water
   _____

   shelter
   _____

   _____

   _____

3. What tool would you use to measure the height of a plant?

   thermometer   hand lens   ruler   scale

4. Which animal would find shelter in a cave?

   deer   beaver   bear   camel

5. Write T if the statement is true or F if it is false.

   _____ Plants need nutrients to survive.

6 – 7.   Look at the words below. Put each word under the correct heading.

cloud   oak tree   rock   milk   dog

Living	Nonliving

A **microscope** is a tool scientists use to magnify an object. Microscopes are helpful to see objects that are too small to see with only your eyes. You might use a microscope to see tiny living things in a sample of water from a pond.

8.   Name a kind of **weather**.

_____

Lesson #15

Mammals

Animals are grouped by what their body looks like and by how or where they live. Most **mammals** have hair or fur on their bodies. Mammal babies usually grow inside the mother and are born alive. Mothers

also make milk for their babies to drink. Mammals use lungs to breathe. Some mammals are unusual. Dolphins and whales are mammals that don't have hair. They live in the ocean and breathe with lungs. A bat is also a mammal, even though it has wings and flies, like a bird.

1. A _____ has hair or fur on its body.

2. Write **T** if the statement is true or **F** if it is false.

_____ Mammals use gills to breathe.

_____ Dolphins and whales are mammals.

3. Which type of cloud is shown?

cirrus   cumulus   stratus

4. Which is **not** something that most plants need to live and grow?

water   air   shelter   nutrients

5. Which tool is shown?

_____

6. What do these animals have in common?

cat   squirrel   whale   deer

A) They are all nonliving.
B) They all have feet.
C) They are all mammals.

7. Which of these is **nonliving**?

butterfly   stone   tulip

8. Which season follows spring? _____

Lesson #16

1. Which of these is a **mammal**?

   A)    B)    C)

2. Write **T** if the statement is true or **F** if it is false.

   _____ Most mammals make milk for their babies.

3. Which type of cloud is shown?

   cirrus   cumulus   stratus

4. Plants have **four** basic needs. Fill in the ones that are missing.

   <u>light_____</u>

   <u>nutrients_____</u>

   _____

   _____

5. Why do animals need shelter?

       for fun   for water   for safety

6. Which step comes after **do an experiment**?
   (See Lesson #9.)

   A) Observe and ask questions.

   B) Share what you learned.

   C) Make a guess.

7. What do you call a tool that is used to magnify, or make something look larger?

      weather vane   hand lens   thermometer

8. For each thing in the chart below, decide whether it is **living** or **nonliving** and put a ✓ in the correct column.

	Living	Nonliving
pine tree		
air		
horse		

Lesson #17

Birds

Birds are another group of animals. They have feathers and wings. Birds breathe with lungs. They lay eggs with hard shells. Birds take care of their babies until they can get their own food.

1. _____ have feathers and wings.

2. Birds lay _____.

3. _____ have hair or fur on their bodies.

4. What are the **four** basic needs of animals?

    A) food, air, wings, water

    B) food, water, shelter, air

    C) sun, food, water, shelter

    D) air, plants, sun, shelter

5. Which of these does **not** describe adult mammals?

A) give birth to live young

B) have hair or fur

C) breathe with gills

D) feed milk to their young

6. What do these animals have in common?

   owl   sparrow   penguin   duck

A) They all live in the water.

B) They are all birds.

C) They all have hair or fur.

7. Nutrients come from the _____.

   air   water   soil

8. Which of these are traits of **birds**?

A) lay eggs and have feathers

B) have gills and live in the water

C) have hair or fur and produce milk for their young

Lesson #18

Fish

Fish are a group of animals that live in water and breathe through gills. Scales cover their bodies. Most fish have fins to help them swim fast. Fish lay eggs.

1. _____ live in water and breathe with gills.

2. Which tool is helpful to see objects that are too small to see with your eyes alone?

    hand lens   thermometer   microscope

3. Put these steps in the correct order. Number from 1 – 4. (See Lesson #9 for help.)

_____ Do an experiment.

_____ Observe and ask questions.

_____ Share what you learned.

_____ Make a guess.

4 – 5.   Match each type of cloud to its description.

_____   They look like wisps of hair.

A) cumulus

_____   They look puffy.

B) cirrus

_____   They look like a sheet.

C) stratus

6. Which is **not** a trait of fish?

   A) breathes with gills   C) lays eggs

   B) has fur   D) lives in water

7. _____ is what the air outside is like.

       Wind   Weather   Seasons

8. All living things have **three** needs. List them.

   _____

   _____

   _____

Lesson #19

1. Which tool would you use to look more closely at a cricket in a jar?

    thermometer   hand lens   microscope

2. For each thing in the chart below, decide whether it is **living** or **nonliving** and put a ✓ in the correct column.

	Living	Nonliving
daisy		
snow		
sand		
squirrel		

3. Draw a cumulus cloud in the box below.

4. Which type of severe weather is shown?

thunderstorm   tornado   blizzard

5. _____ have feathers and wings.

6. Animals have **four** basic needs. Fill in the ones that are missing.

   air _____   food _____

   _____   _____

7. Write T if the statement is true or
   F if it is false.

   _____ Fish breathe with lungs.

8. How are these animals alike?

         bear   bat   skunk   tiger

   A) They all walk.
   B) They all breathe with gills.
   C) They are all mammals.

Lesson #20

Reptiles

Reptiles have skin with dry scales. They use lungs to breathe. Most reptiles have four legs. Most reptiles lay eggs. A turtle is the only reptile that has a shell. A snake is a reptile. Lizards and alligators are reptiles, too.

1. _____ have skin with dry scales.

2. What is the only reptile that has a shell?

   _____

3. Give an example of a **reptile**. _____

4. Which tool is shown?

   _____

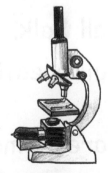

Use the chart to answer the questions below.

	Breathe	Outside Covering	Babies
Mammals	lungs	hair or fur	born
Fish	gills	scales	hatched from eggs
Birds	lungs	feathers	hatched from eggs

5. Which of these animal groups lay eggs?

   A) birds and mammals

   B) birds and fish

   C) fish and mammals

6. Which group breathes with gills?   _____

7. Which group has hair or fur?   _____

8. Which groups breathe with lungs?

   _____

   _____

Lesson #21

Amphibians

Amphibians live part of their lives in the water and the other part on land. They have wet, smooth skin. Amphibians lay their eggs in water. When they are babies, they use gills to breathe. When they are adults, they use lungs to breathe. Frogs, salamanders, and toads are amphibians.

1. _____ have wet, smooth skin.

2. When they are adults, amphibians use

_____ to breathe.

3. Give an example of an **amphibian**.

_____

4. Which type of cloud is shown?

cirrus   cumulus   stratus

5. Match each animal group with its description.

___ mammal   A) has feathers and lays eggs

___ reptile   B) has hair or fur and gives
                     birth to live young

___ fish   C) has skin with dry scales

___ bird   D) lays eggs and breathes
                     with gills

6. Which tool would you use to see something too small to
   see with only your eyes?

   thermometer   hand lens   microscope

7. When rivers and lakes overflow, what can happen?

   a tornado   a drought   a flood

8. Which step comes after **observe and ask questions**?

   A) Do an experiment.
   B) Share what you learned.
   C) Make a guess.

Lesson #22

Use the chart to answer the questions below.

	Breathe	Outside Covering	Babies
Mammals	lungs	hair or fur	born alive
Fish	gills	scales	hatched from eggs
Birds	lungs	feathers	hatched from eggs
Reptiles	lungs	dry scales	hatched from eggs
Amphibians	gills/lungs	wet and moist	hatched from eggs

1. How are amphibians like fish when they are first born?

A) They both have scales.

B) They both breathe with gills.

C) They both breathe with lungs.

2. Which animal group has wet and moist skin?

_____

3. A squirrel is an example of a(n) _____.

    bird   fish   mammal   amphibian

4. Plants have **four** basic needs. Fill in the ones that are missing.

    _water_____

    _____

    _____

    _nutrients_____

5. Which of these is **nonliving**?

    bumblebee   wind   lion

6. Which tool measures temperature?

    thermometer   hand lens   weather vane

7. Write T if the statement is true or F if it is false.

    _____ Mammals make milk for their young.

8. Which season follows summer? _____

Lesson #23

1. How are reptiles and mammals alike?

A) Both have feathers.

B) Both breathe with lungs.

C) Both have hair or fur.

2. All living things need _____.

A) air, food, and water

B) food, wings, and water

C) air, fur, and water

3. Name **two** examples of **weather**.

_____

_____

4. Which type of cloud looks like wisps of hair?

    cirrus   cumulus   stratus

5. _____ is what the air outside is like.

    Thermometer   Seasons   Weather

6 – 7.   For each living thing in the chart below, decide whether it is **a mammal**, **a bird**, or **a reptile**, and put a ✓ in the correct column.

	Mammal	Bird	Reptile
snake			
penguin			
raccoon			
chicken			
dog			
lizard			

8.   What is it called when the land gets dry and hard because it has not rained for a very long time?

   A) flood

   B) tornado

   C) drought

   D) blizzard

Lesson #24

Life Cycles of Animals (Part 1)

Every animal has a life cycle. When an animal has a baby, a life cycle begins. A **life cycle** is all of the parts of an animal's life. When most baby animals are born, they look just like their parents, only smaller.

1. A _____ is all the parts of an animal's life.

2. Which type of cloud is shown?

   cirrus   cumulus   stratus

3. Write T if the statement is true or F if it is false.

   _____ All living things need food, soil, and shelter.

4. Which animal group gives birth to live young?

_____

5. An animal that has feathers and wings is a _____.

        fish   reptile   bird   mammal

6. Put these steps in the correct order. Number from 1 – 4.
    (See Lesson #9 for help.)

    _____   Share what you learned.

    _____   Do an experiment.

    _____   Make a guess.

    _____   Observe and ask questions.

7. Draw a picture of a **living** thing.

8. Name this tool.

_____

Lesson #25

Life Cycles of Animals (Part 2)

A frog has an unusual life cycle. A young frog that hatches from an egg is called a **tadpole**. Tadpoles live in water. Tadpoles use gills to breathe. As the tadpole gets bigger, it grows two back feet and then two front feet. Soon the tadpole begins to look more like a frog, but it still has a tail. Once the frog is fully grown, it loses its tail and uses its lungs to breathe. It can now live on land. It now looks like its parents.

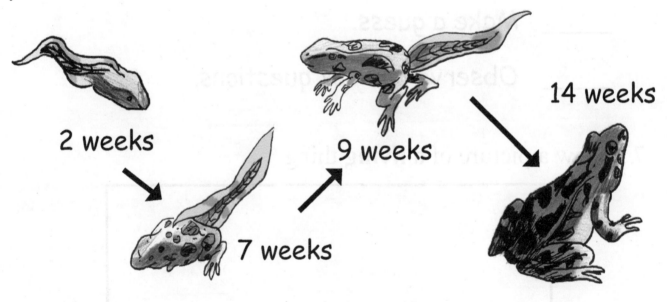

1. A young frog is called a _____.

2. Tadpoles use _____ to breathe.

3. Frogs belong to which animal group?

mammal   reptile   bird   amphibian

4 – 5.   Fill in the chart about animal groups. Some have
        been done for you. (See Lesson #22 for help.)

	Breathe	Outside Covering
Mammals	lungs	
Fish		scales
Birds	lungs	
Reptiles		dry scales
Amphibians	gills/lungs	

6. A weather vane measures _____.

    wind direction   temperature   wind speed

7. Which animal would find shelter in or near the water?

        squirrel   beaver   cat   camel

8. All parts of an animal's life are called the _____.

        shelter   life cycle   group

Lesson #26

1. Animals have four basic needs. Fill in the ones that are missing.

water _____

_____

air _____

_____

2. Tadpoles use _____ to breathe.

3. How are these animals alike?

      rabbit   lizard   frog   cardinal

A) They all fly.

B) They are all mammals.

C) They all breathe with lungs.

4. Which tool would you use to take a closer look at a butterfly wing?

     weather vane   thermometer   hand lens

5. Write **T** if the statement is true or **F** if it is fa

_____ Living things need food, water, and a

6. Which type of cloud is shown?

cirrus   cumulus   stratus

7. Underline the **living** things.

mouse   snow   oak tree   wind

8. For each living thing in the chart below, decide whether it is a **mammal**, an **amphibian**, or a **reptile** and put a ✓ in the correct column.

	Mammal	Amphibian	Reptile
lizard			
squirrel			
frog			
turtle			

sson #27

ed water, light, air, and
of the plant helps the plant get
are three main parts to every
eaf **(leaves)**, and the **stem**.

st plants grow underground.
Roots ter and nutrients from the soil.

- The **stem** grows above the ground and it helps to hold the plant up. The stem carries water and nutrients from the roots to the leaves.

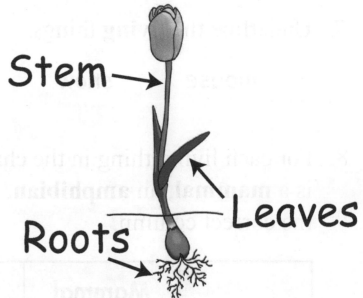

- The **leaves** grow out of the stem. This is where the plant makes its food. **Air, water**, and **sunlight** are all needed to make food that the plant needs to live.

1. Which plant part grows underground? _____

2. Which part holds the plant up? _____

3. Which part of the plant makes food? _____

4. All parts of an animal's life are called the _____.

        shelter   life cycle   group

5 – 6. Put these steps in the correct order. Number from 1 – 4.

_____ Make a guess.

_____ Do an experiment.

_____ Observe and ask questions.

_____ Share what you learned.

7. An animal that lives in the water and breathes with gills is a _____.

        fish   reptile   bird   mammal

8. _____ is what the air outside is like.

        Wind   Weather   Seasons

Lesson #28

Animal Changes

It is hard for some animals to find food in the winter. Some animals go into a deep sleep for the winter. This is called **hibernation**. Some

animals **migrate** (move to a warmer place) for the winter in order to find food. These animals won't be able to find food once the ground is covered with snow and the lakes and ponds are frozen.

1. Some animals go into a deep sleep for the winter. What is this called?

   life cycle   hibernation   migration

2. Some animals move to a warmer place during the winter so they can find food. What is this called?

   life cycle   hibernation   migration

3. Which season follows fall (autumn)?

   _____

4. To which animal group
   does this animal belong?

   A) reptile

   B) mammal

   C) amphibian

   D) bird

5. Which are **nonliving** things?

   tree   soil   bumblebee   sun

6. Plants have **four** basic needs. List them.

   _____   _____

   _____   _____

7. Which animal group gives birth to live young?

   reptile   amphibian   mammal

8. Which type of cloud is shown?

   cirrus   cumulus   stratus

Lesson #29

Life Cycle of a Plant

Just like animals, plants have a **life cycle**. All the stages of a plant's life make up its life cycle. The life cycle begins with a **seed**. With water and sunlight, the seed begins to grow. Then, the plant grows into an adult plant. Finally, it makes seeds that will grow into new plants.

1. The plant life cycle begins with a _____.

2. Match each plant part with its definition.

   ___ leaf   A) holds the plant up

   ___ stem   B) where the plant makes its food

   ___ root   C) takes in water and nutrients from the soil

3. A young frog is called a _____.

4. Name this tool.

_____

5. When rivers and lakes overflow, what can happen?

   a tornado   a drought   a flood

6. Write T if the statement is true or F if it is false.

   _____   Living things make new living things that are like themselves.

7. Nutrients come from the _____.

   air   water   soil

8. Match each animal group with the words that describe it.

   ____ bird   A) has dry, scaly skin and lays eggs

   ____ reptile   B) has wings, feathers, and lays eggs

   ____ amphibian   C) has moist skin and stays close to water

Lesson #30

Soil

Mostlants need soil to grow. The soil also helps to hold the roots of the plant in place. **Soil** is made up of small pieces of rock and leftover bits of dead  plants and animals. There are many kinds of soil. Soil can be different colors; some can hold more water than others, and some have more rock pieces in them. Some soil has **clay** in it. Another type of soil is called **silt**. Silt is smoother than clay; it feels like powder. **Sand** is another type of soil. Sand has tiny grains of rock that you can easily see with your eye. Different soils are good for growing different kinds of plants.

1. Name **two** kinds of soil.

    _____

    _____

2. The plant life cycle begins with a _____.

3 – 4. Look at the diagram.
        Label the parts of the plant.

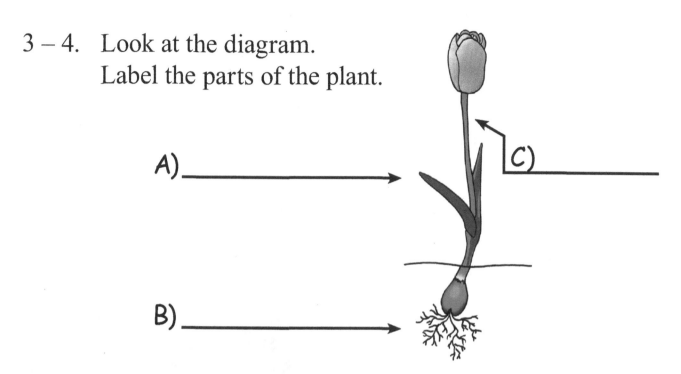

A)_____

C)_____

B)_____

5. Some animals go into a deep sleep for the winter. What is this called?

     hibernation   life cycle   migration

6. An animal that has wet, moist skin and begins its life in the water is a _____.

     reptile   amphibian   mammal

7. Write **T** if the statement is true or **F** if it is false.

     _____ Living things need food, water, and air.

8. Which tool measures temperature?

     thermometer   hand lens   ruler   scale

Lesson #31

1. Some animals move to a warmer place during the winter so they can find food. What is this called?

    life cycle   migration   hibernation

2. Put these steps in the correct order. Number from 1 – 4.

    _____ Do an experiment.

    _____ Observe and ask questions.

    _____ Share what you learned.

    _____ Make a guess.

3. All parts of an animal's life are called the _____.

    shelter   hibernation   life cycle

4. What are the **four** basic needs of animals?

    A) food, plants, feet, water

    B) food, water, shelter, air

    C) sun, food, water, shelter

5. Which part holds the plant up?

        leaf   root   stem

6. Why do animals need shelter?

      for protection   for food   for oxygen

7. Match each animal group to its definition.

____ bird	A)	has hair or fur; feeds young with milk
____ reptile	B)	two legs; wings and feathers
____ fish	C)	moist skin, lives near water
____ mammal	D)	dry, scaly animal that lays eggs
____ amphibian	E)	lives whole life in water; breathes with gills

8. This picture shows what happens when there is a _____.

  flood   drought   blizzard

Lesson #32

Stars

A **star** is a huge ball of glowing gas. These hot gases give off heat and light. The sun is a star. It is the closest star to Earth. The sun is a medium-sized star. We see the sun in the daytime, but all other stars are seen in the night sky. Stars look like tiny dots in the sky because they are

Big Dipper and Little Dipper

so far away from us. A group of stars that form a pattern are called **constellations**. The Big Dipper and The Little Dipper are examples of constellations.

1. A _____ is a huge ball of glowing gas.

2. The Big Dipper is an example of a _____.

      planet   moon   constellation

3. Which part of the plant makes food?

    A) leaf

    B) stem

    C) root

4. Write **T** if the statement is true or **F** if it is false.

_____   The sun is a star.

5. Which tool measures wind direction?

thermometer   hand lens   weather vane

6. Plants have **four** basic needs. One has been done for you. List the other three.

light _____

_____

_____

_____

7. A nest may be a kind of _____ for a bird.

shelter   tool   food

8. Which plant part grows underground?

leaf   stem   root

Lesson #33

The Moon

The **moon** is a large ball of rock that orbits (travels around) the Earth. It takes the moon about one month to travel around Earth. There is no air on the moon. The moon reflects light from the sun. The moon is always moving, so the lighted part you see from Earth changes each night (see pictures to the right). These changes follow a pattern that repeats once a month. These changes are called **phases**.

1. The _____ is a large ball of rock that orbits the Earth.

2. About how long does it take the moon to orbit the Earth?

    a day   a year   a month

3. Which of these is **nonliving**?

   A) daisy

   B) moon

   C) moose

4. Some animals move to a warmer place during the winter so they can find food. What is this called?

   migration   life cycle   hibernation

5. Which of these do all living things need?

   feet   wings   food   feathers

6. Owls belong to which animal group?

   mammal   reptile   bird   amphibian

7. The three main parts of a plant are the roots, leaves, and _____.

   A) nutrients
   B) stem
   C) water

8. List **two** examples of a **mammal**.

   _____

   _____

Lesson #34

1. Match each plant part with its definition.

    ___ stem   A) holds the plant up

    ___ root   B) where the plant makes its
                         food

    ___ leaf   C) takes in water and
                         nutrients from the soil

2. The Little Dipper is an example of a _____.

    planet   constellation   moon

3. The sun is a _____.

4. Some animals go into a deep sleep for the winter. What is
    this called?

    life cycle   hibernation   migration

5. Underline the **mammals**.

    turkey   rabbit   bear   robin

    frog   snake   cat   catfish

    lizard   raccoon   owl   turtle

6. About how long does it take the moon to orbit the Earth?

   a month   a year   a day

7. Name these tools.

   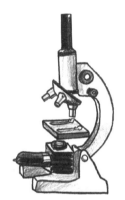

A) _____   B) _____

8. For each thing in the chart below, decide whether it is
   living or **nonliving** and put a ✓ in the correct column.

	Living	Nonliving
moon		
goat		
sunflower		

Lesson #35

1. Which part holds up the plant?

    leaf   root   stem

2. Sand, silt, and clay are types of _____.

    weather   toys   soil   clouds

3. Which type of cloud is shown?

    cirrus   cumulus   stratus

4. Some animals move to a warmer place during the winter so they can find food. What is this called?

    life cycle   hibernation   migration

5. Turtles belong to which animal group?

    mammal   reptile   bird   amphibian

6. Write T if the statement is true or F if it is false.

    _____   Stars that form a pattern are called constellations.

7. Match each animal group to its definition.

_____ fish

_____ reptile

_____ mammal

_____ amphibian

_____ bird

A) has hair or fur; feeds young with milk

B) two legs; wings and feathers

C) moist-skin; lives near water

D) dry, scaly animal that lays eggs

E) lives whole life in water; breathes with gills

8. Animals have **four** basic needs. One has been done for you. List the other three.

<u>water</u>

_____

_____

_____

Lesson #36

1. Which plant part grows underground?

    leaf   stem   root

2. Put these steps in the correct order.

    _____ Make a guess.

    _____ Share what you learned.

    _____ Do an experiment.

    _____ Observe and ask questions.

3. A weather vane measures _____.

    wind direction   temperature   wind speed

4. _____ is what the air outside is like.

    Wind   Weather   Seasons

5. An animal that has scales and breathes with gills
   is a _____.

    fish   reptile   bird   mammal

6. Match each word with its definition.

    ____ wind   A) no rain for long periods

    ____ drought   B) what the outside air is like

    ____ flood   C) moving air

    ____ weather   D) when rivers and lakes
                             overflow onto the land

7. How are these animals alike?

    squirrel   snake   toad   penguin

    A) They all live in the water.

    B) They all breathe with lungs.

    C) They are all reptiles.

8. Look at the diagram.
    Label the parts of the plant.

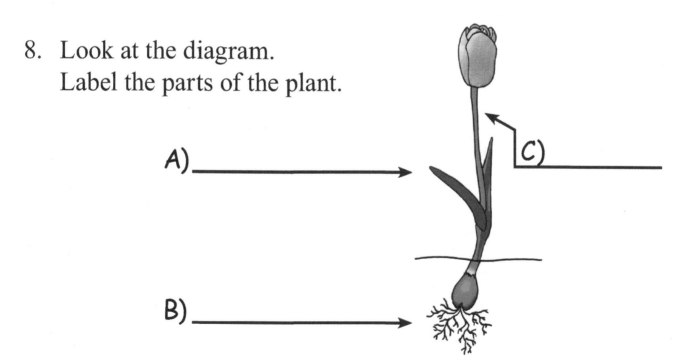

A) _____

C) _____

B) _____

Lesson #37

What is a Solid?

A solid is a form of matter. **Matter** is anything that takes up space. Matter is all around you. The things you can see, and even the air you can't see, are matter. A **solid** has its own shape and takes up a certain amount of space. Your desk, a pencil, a

banana, and a car are all examples of solids. Solids come in different sizes, shapes, and colors. A solid will stay the same unless you break it or change it in some other way.

1. A _____ has its own shape and takes up a certain amount of space.

2. Give an example of a **solid**. _____

3. The plant life cycle begins with a _____.

4. An animal that has dry skin with scales and lays eggs is a _____.

       reptile   amphibian   mammal

5. Color the part of the plant that takes in water from the soil.

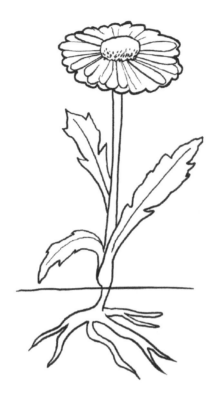

6. About how long does it take the moon to orbit the Earth?

    a month   a year   a day

7. The sun is a _____.

    planet   constellation   star

8. Sand, silt, and clay are types of _____.

    weather   insects   soil   clouds

Lesson #38

1. A hole in the ground may be a kind of _____ for a groundhog.

   tool   food   shelter

2. Color the part of the plant that holds the plant up.

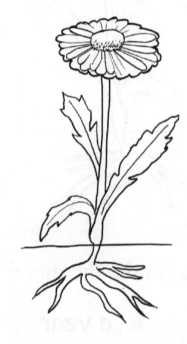

3. To which group does this animal belong?

   A) reptile

   B) mammal

   C) amphibian

   D) bird

4. A young frog is called a _____.

5. Match each animal group to its definition.

    ___ bird   A) has hair or fur; gives birth to live young

    ___ amphibian   B) two legs; wings and feathers

    ___ fish   C) moist-skin; lives near water

    ___ mammal   D) animal with dry, scaly skin

    ___ reptile   E) has scales and breathes with gills

6. All parts of an animal's life are called the _____.

   A) shelter
   B) life cycle
   C) group

7. Give an example of a **solid**. _____

8. Write T if the statement is true or F if it is false.

       _____ All living things need food, water, and air.

Lesson #39

What is a Liquid?

A liquid is also a form of matter. A liquid does not have its own shape. A **liquid** takes the shape of its container. Juice, milk, and water are all examples of liquids. In the examples to the right, the liquid takes the shape of its container. The milk in the carton has the shape of the carton. If you

poured some milk from the carton into a glass, the milk would take the shape of the glass.

1. A _____ takes the shape of its container.

2. Give **two** examples of a **liquid**.

_____

_____

3. Which season follows spring? _____

4. Which type of cloud looks like cotton?

    cirrus   stratus   cumulus

5. Animals have four basic needs. List them.

    _____

    _____

    _____

    _____

6. Which type of matter has its own shape and takes up a certain amount of space?

                liquid   solid

7. The Big Dipper is an example of a _____.

        planet   constellation   moon

8. List an example of a **mammal**.

    _____

Lesson #40

1. A _____ can happen when there is too much rain.

   A) drought

   B) tornado

   C) flood

2 – 3. For each thing in the chart below, decide whether it is **living** or **nonliving** and put a ✓ in the correct column.

	Living	Nonliving
rose bush		
mouse		
rock		
sun		

4. Which is **not** true of amphibians?

   A) breathes with lungs

   B) has moist skin

   C) lays eggs

   D) has hair or fur

5. Which tool would you use to see something too small to see with only your eyes?

   thermometer   microscope   hand lens

6 – 7. For each thing in the chart below, decide whether it is a **solid** or a **liquid** and put a ✓ in the correct column.

	Solid	Liquid
juice		
pencil		
lamp		
rain		

8. Match each plant part with its definition.

   ____ stem   A) holds the plant up

   ____ root   B) where the plant makes its food

   ____ leaf   C) takes in nutrients from the soil

Lesson #41

What is a Gas?

A gas is also a form of matter. A **gas** does not have its own shape. A gas fills the shape of its container. The air you breathe is made up of gases.

When you blow into a balloon, the air fills the balloon. The air takes the shape of the balloon. Usually, you can't see or smell the air around you, but you can see what it does. It can blow your hair, help a kite fly, and make leaves in the trees move back and forth. Oxygen and helium are types of gases.

1. A _____ fills the shape of its container.

2. Give an example of a **solid**. _____

3. Tadpoles breathe with _____.

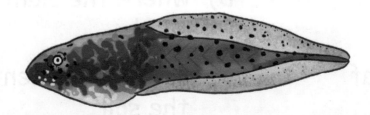

4 – 5. For each thing in the chart below, decide whether it is a **solid**, a **liquid**, or a **gas** and put a ✓ in the correct column.

	Solid	Liquid	Gas
milk			
notebook			
air			

6. Write T if the statement is true or F if it is false.

_____ A thermometer tells how hot or cold the air is.

7. _____ have feathers and wings.

8. A plant's nutrients come from the _____.

air   water   soil

Lesson #42

1. Color the part of the plant where food is made.

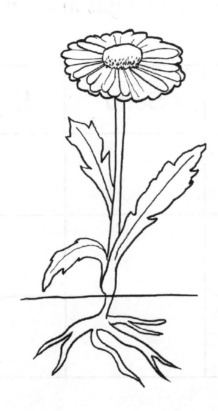

2. Name this tool.

   _____

3. Underline the **mammals**.

   cardinal   raccoon

   monkey   frog

   alligator   horse

   kitten   catfish

4. Write T if the statement is true or F if it is false.

_____ Animals need shelter for protection.

5. Which of these is a **living** thing?

cloud   oak tree   sand

6. All parts of an animal's life are called the ___.

shelter   life cycle   group

7. Which type of cloud is shown?

cirrus   cumulus   stratus

8. Plants have **four** basic needs. List them.

_____

_____

_____

_____

Lesson #43

Force

A **force** is a push or a pull. A push moves an object away from you, and a pull moves an object toward you. When you push your mom in a wagon, you are using a pushing force. When you pull something in a wagon,

you are using a pulling force. Whenever you change the direction of a moving object, you need force. If you use more force, you can move an object faster and farther.

1. A _____ is a push or pull.

2. The plant life cycle begins with a _____.

   A) bean

   B) seed

   C) pot

3. Write T if the statement is true or F if it is false.

   _____ Living things need food, water, and air.

4. Underline **two** kinds of **soil**.

sand   pebble   clay   powder

5. Look at the diagram. Label the parts of the plant.

A)_____

B)_____

C)_____

6. Some animals move to a warmer place during the winter so they can find food. What is this called?

life cycle   migration   hibernation

7. A _____ is a huge ball of glowing gas.

planet   star   moon

8. About how long does it take the moon to orbit the Earth?

a day   a year   a month

Lesson #44

1. Stars that form a pattern are called _____.

    constellations   planets   pictures

2. Put these steps in the correct order. Number from 1 – 4.

    _____ Do an experiment.

    _____ Observe and ask questions.

    _____ Share what you learned.

    _____ Make a guess.

3. What does this tool measure?

    A) temperature
    B) wind speed
    C) wind direction

4. A _____ is a push or pull.

5. Which animal group gives birth to live young?

    reptiles   birds   mammals   fish

6 – 7. Put each word in the correct column of the graphic organizer.

oxygen   orange juice   lamp   helium

rain   dish soap   paper   bike

Solid	Liquid	Gas

8. Some animals go into a deep sleep for the winter. What is this called?

A) hibernation

B) life cycle

C) migration

Lesson #45

Gravity

Gravity is a force that pulls objects toward each other. When you throw a ball into the air, you know it will come back down. The ball comes back to you because Earth's gravity pulls on it. Earth's gravity pulls objects toward the center of the Earth.

Gravity acts on objects without even touching them.

1. The force that pulls objects toward each other is called _____.

   A) force

   B) constellation

   C) gravity

2. Which type of cloud is shown?

   cirrus   cumulus   stratus

3. A _____ is a push or pull.

4. _____ is what the air outside is like.

    Thermometer   Seasons   Weather

5 – 6. Match each animal group with its description.

    ____ mammal   A) has feathers and lays eggs

    ____ reptile   B) has hair or fur and gives
                            birth to live young

    ____ fish   C) has skin with dry scales

    ____ bird   D) lays eggs and breathes
                            with gills

    ____ amphibian   E) has wet skin; lives part of
                            its life in the water and
                            part on land

7. Which tool would you use to see something too small to
    see with only your eyes?

    microscope   hand lens   thermometer

8. Which plant part grows underground?

        leaf   stem   root

Lesson #46

1. Match each plant part with its definition.

___ root   A) holds the plant up

___ stem   B) where the plant makes its food

___ leaf   C) takes in water and nutrients from the soil

2. The force that pulls objects toward each other is called _____.

    life cycle   gravity   constellation

3. Which of these is **living**?

snow   daisy   rocks

4. How are reptiles and mammals alike?

A) Both breathe with lungs.

B) Both have hair or fur.

C) Both lay eggs.

5. The plant life cycle begins with a _____.

bean   seed   stem

6 – 7.   For each living thing in the chart below, decide whether it is a **mammal**, an **amphibian**, or a **reptile** and put a ✓ in the correct column.

	Mammal	Amphibian	Reptile
alligator			
chipmunk			
toad			
turtle			
frog			
monkey			

8.   Some animals move to a warmer place during the winter so they can find food. What is this called?

A) constellation

B) hibernation

C) migration

Lesson #47

What Makes Sound?

Sound is a form of energy that you can hear. Sound travels in waves. You hear a dog bark, a horn honk, or a whistle blow. All of these sounds come from objects that **vibrate**, or move back and forth. You hear sounds when vibrations move through the air to your ears.

1. Sound travels in _____.

2. The word that means "to move back and forth" is

   _____.

3. Give an example of something that makes a sound.

   _____

4. A _____ is a push or pull.

5 – 6. Put each word in the correct column of the graphic organizer.

notebook   soda pop   oxygen

milk   glass   juice

Solid	Liquid	Gas

7. Which part of a plant grows underground?

leaf   stem   root

8. About how long does it take the moon to orbit the Earth?

a day   a year   a month

Lesson #48

1. Write **T** if the statement is true or **F** if it is false.

   _____   Sound comes from objects that vibrate.

2. Animals have **four** basic needs. One has been done for you. List the other three.

   _shelter_____

   _____

   _____

   _____

3. Which part holds the plant up?

   leaf   stem   root

4. Stars that form a pattern are called _____.

   planets   pictures   constellations

5. A hive may be a kind of _____ for a bee.

   tool   shelter   food

6. Look at the diagram.
   Label the parts of the plant.

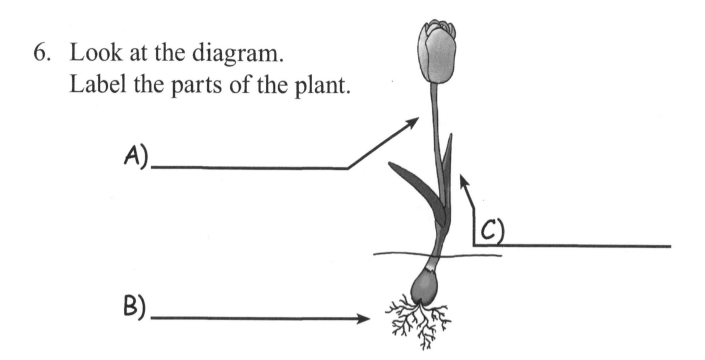

A)_____

B)_____

C)_____

7. How can you tell it is spring?

   A) Some leaves change color.

   B) Trees have no leaves.

   C) Trees grow new leaves.

8. Put these steps in the correct order. Number from 1 – 4.

   _____ Make a guess.

   _____ Observe and ask questions.

   _____ Share what you learned.

   _____ Do an experiment.

Lesson #49

Natural Resources (Part 1)

A **natural resource** is something that is found in nature that people need or can use. **Water** is

a natural resource. People use water to cook, bathe, water plants, swim, and drink. **Air** is another natural resource. People need air to breathe. Moving air can also be used as a source of energy. Two other natural resources are **rocks** and **soil**. Rocks give us metals we need to make things like cars and computers. People use soil to grow plants.

1. List **two** examples of a **natural resource**.

   _____

   _____

2. All living things need air, food,

   and _____.

3. Something that is found in nature that people need or can use is called a(n) _____.

   life cycle   natural resource   force

4. The word that means "to move back and forth" is _____.

   constellation   vibrate   energy

5. The force that pulls objects toward each other is called _____.

   vibration   gravity   constellation

6. A _____ is a huge ball of glowing gas.

   planet   star   moon

7. Sand, silt, and clay are types of _____.

   weather   toys   soil   clouds

8. Give an example of a **liquid**. _____

Lesson #50

Natural Resources (Part 2)

Plants and **animals** are also natural resources. People use plants and animals as food. Plants also give off oxygen. We need oxygen to stay alive. Wood

cotton plant

from trees helps to make furniture, paper, and the houses we live in. The cotton used to make clothes comes from the cotton plant. Animals also give us fur for clothing.

1. Name **two** natural resources from the teaching above.

    _____

    _____

2. A _____ is a push or pull.

3. Which animal group lives its entire life in the water?

    A) reptile   C) amphibian

    B) bird   D) fish

4. Which type of cloud is shown?

cirrus   cumulus   stratus

5. The plant life cycle begins with a _____.

bean   stem   seed

6. Match each plant part with its definition.

____ stem   A) takes in water and
                      nutrients from the soil

____ root   B) where the plant makes its
                      food

____ leaf   C) holds the plant up

7. Which of these is a **gas**?

paint   glass   oxygen   water

8. To which group does this animal belong?

A) reptile
B) mammal
C) amphibian
D) bird

Lesson #51

1 – 2. For each living thing in the chart below, decide whether it is a **fish**, a **bird**, or a **reptile** and put a ✓ in the correct column.

	Fish	Bird	Reptile
turkey			
salmon			
penguin			
turtle			
alligator			
trout			

3. Underline the **liquids**.

glue   juice   cotton

cloud   milk   stick

4. Which season follows summer? _____

5. Which tool is used to find the direction of the wind?

weather vane   thermometer   hand lens

6. Color the part of the plant that takes in water from the soil.

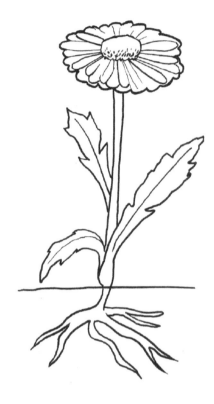

7. Write **T** if the statement is true or **F** if it is false.

_____ Mammals make milk for their young.

8. Plants have four basic needs. List them.

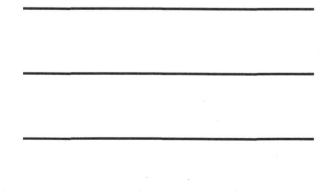

Lesson #52

Harming Natural Resources

Waste that harms the water, air, or land is called **pollution**. Pollution can kill plants and make people and animals sick. Pollution comes in many forms. **Land pollution** is caused by people littering. **Water pollution** comes from trash, waste from

factories, or oil spills. **Air pollution** comes from factory and car fumes. Stopping pollution will save our land, air, and water.

1. Waste that harms the water, air, or land is called

_____.

2 – 3. Name **three** kinds of pollution.

_____

_____

_____

4. This picture shows a _____.

flood   drought   tornado

5. Which animal group spends part of its life in the water and part on land?

reptile   bird   amphibian   fish

6. All parts of an animal's life are called the _____.

life cycle   resource   shelter

7. Put these steps in the correct order.

_____ Make a guess.

_____ Share what you learned.

_____ Do an experiment.

_____ Observe and ask questions.

8. About how long does it take the moon to orbit the Earth?

a year   a day   a month

Lesson #53

1. Match each animal with its shelter.

squirrel   underground

bat   tree

rabbit   cave

2. Name this tool.

_____

3. All living things need water, food, and _____.

4. Waste that harms the water, air, or land is called _____.

natural resource   pollution   drought

5. Underline two natural resources.

car   water   plants

6. Which of these does **not** describe adult mammals?

A) breathes with gills

B) gives birth to live young

C) feeds milk to their young

D) has hair or fur

7. Match each type of cloud to its description.

A) stratus

_____ They look like wisps of hair.

B) cirrus

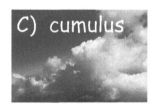

_____ They look puffy.

C) cumulus

_____ They look like a sheet.

8. Which tool would you use to look more closely at a butterfly cocoon?

thermometer   hand lens   microscope

Lesson #54

1. What kind of pollution comes from factories and cars?

    land pollution   air pollution   water pollution

2. What kind of pollution is caused by people littering?

    land pollution   air pollution   water pollution

3. Underline the **liquids**.

    paint   water   stick   pan   shampoo

4. Draw a cumulus cloud in the box.

5. Write **T** if the statement is true or **F** if it is false.

    _____ Fish breathe with gills.

6. Which tool is shown?

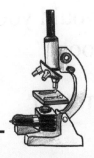

    _____

7. Animals have **four** basic needs. List them.

_____

_____

_____

_____

This symbol means that a product can be recycled. When you **recycle**, you are helping to save resources by reusing them. When an item is recycled, it is broken down or changed in some way so that the material can be used to make something new. Some materials that can be recycled are **paper**, **aluminum cans**, some **plastics**, and **glass**.

8. Name **two** materials that can be recycled.

_____

_____

Lesson #55

1. Which plant part grows underground?

    leaf   stem   root

2. According to Lesson #52, what kind of pollution comes from oil spills?

    land pollution   air pollution   water pollution

3. What does this symbol mean?

    _____

4. To which group does this animal belong?

    A) reptile

    B) mammal

    C) amphibian

    D) bird

5. A _____ is a huge ball of glowing gas.

    A) star

    B) moon

    C) planet

6. Waste that harms the water, air, or land is called _____.

   A) pollution

   B) natural resource

   C) life cycle

7. For each thing in the chart below, decide whether it is **living** or **nonliving** and put a ✓ in the correct column.

	Living	Nonliving
drinking straw		
grasshopper		
oak tree		
chair		

8. Which of these animal groups lay eggs?

   A) fish and mammals

   B) birds and mammals

   C) birds and fish

Lesson #56

Word Bank

bird	microscope	stem	thermometer
flood	nonliving	tadpole	weather
mammal	root		

Across

2. a young frog

4. when rivers and lakes get too high and overflow

5. what the air outside is like

6. a tool scientists use to magnify an object

8. part of the plant that holds the plant up

Down

1. this animal has feathers and wings

2. measures the temperature of the air

3. do not need food, water, or air

7. part of the plant that takes in water and nutrients from the soil

9. an animal that has hair or fur on its body

Use the clues and the Word Bank to complete this puzzle.

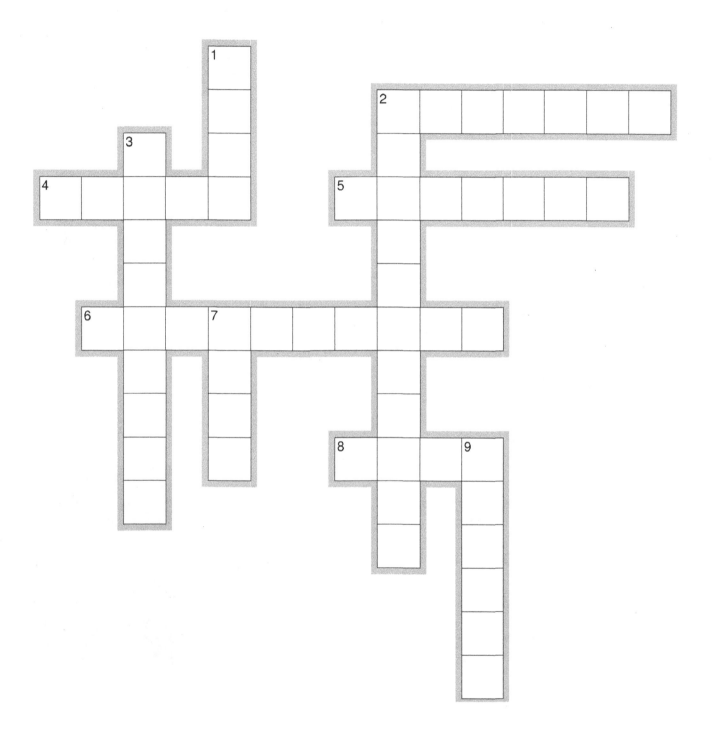

Lesson #57

1. Which part of the plant takes in water from the soil?

    stem   root   leaf

2. Give **two** examples of a **solid**.

    _____

    _____

3. Which tool would you use to see something too small to see with only your eyes?

    thermometer   hand lens   microscope

4. All living things need _____.

    A) air, food, and water

    B) food, wings, and water

    C) air, fur, and water

5. Dry weather that makes the ground hard and cracked is called a _____.

    flood   tornado   drought   blizzard

6 – 7. For each living thing in the chart below, decide whether it is a **mammal**, a **bird**, a **reptile**, or an **amphibian** and put a ✓ in the correct column.

	Mammal	Bird	Reptile	Amphibian
alligator				
turkey				
coyote				
frog				
turtle				
beaver				

8. A weather vane measures _____.

   A) wind direction
   B) temperature
   C) wind speed

Lesson #58

1. What kind of pollution is caused by people littering?

   air pollution   water pollution   land pollution

2. The force that pulls objects toward each other is called _____.

   life cycle   gravity   constellation

3. The Big Dipper is an example of a _____.

   planet   constellation   moon

4. Plants have **four** basic needs. List them.

   _____

   _____

   _____

   _____

5. Which part of the plant makes food?

   root   leaf   stem

6. Which of these can be recycled?

wood   aluminum   glass   paper   brick

7. About how long does it take the moon to orbit the Earth?

a year   a month   a day

8. Match each animal group with the words that describe it.

____ bird   A) has moist skin and stays close to water

____ reptile   B) has wings, feathers, and lays eggs

____ amphibian   C) has hair or fur and makes milk for its young

____ fish   D) has dry, scaly skin and lays eggs

____ mammal   E) spends its whole life in the water; lays eggs

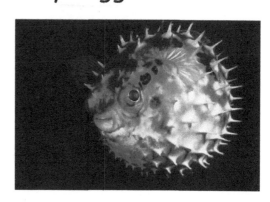

Lesson #59

1. Some animals move to a warmer place during the winter so they can find food. What is this called?

    hibernation   migration   life cycle

2. A _____ is a push or pull.

3. Which type of cloud is shown?

    cirrus   cumulus   stratus

4. Name this tool.

    _____

    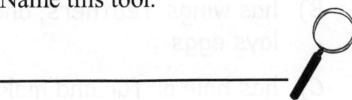

5. Look at the diagram.
    Label the parts of the plant.

    A)_____ →

    B)_____ →

    C)_____

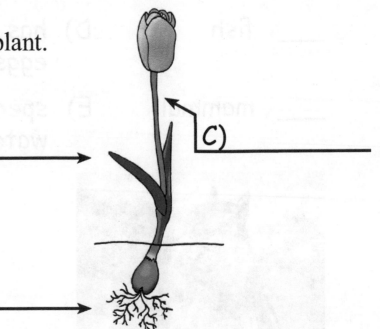

6. Draw the symbol for **recycle** in the box below.

7 – 8. Look at each column. Decide what all the words have
in common. Write the best title at the top of each list.
Use the choices below. One has been done for you.

Reptiles   Mammals   Needs of Animals

Needs of Plants   Stars

Needs of Animals	A)	B)
water	squirrel	air
food	monkey	nutrients
shelter	dog	water
air	lion	sunlight

Lesson #60

1. Waste that harms the water, air, or land is called _____.

   natural resource   life cycle   pollution

2. Which of these is **nonliving**?

   sunflower   sand   giraffe

3. Underline the **liquids**.

   juice   oxygen   apple   water   lemonade

4. Which is **not** a trait of most mammals?

   A) breathes with lungs
   B) gives birth to live young
   C) lays eggs
   D) has hair or fur

5. Which of these **cannot** be recycled?

   A)    B)    C)

6. Which of these measures wind direction?

   thermometer   weather vane   scale

7. Put these steps in the correct order.

_____ Do an experiment.

_____ Observe and ask questions.

_____ Share what you learned.

_____ Make a guess.

8. Look at the graph to answer the questions below.

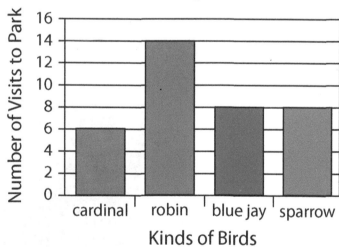

A) Which bird made the fewest visits to the park?

_____

B) Which bird made the most visits to the park?

_____

Lesson #61

1. Which tool is the best instrument to measure the height of a plant's stem?

    A) thermometer

    B) hand lens

    C) ruler

2. Which animal might build a nest in a tree?

    frog   rabbit   cardinal   penguin

3. Write T if the statement is true or F if it is false.

    _____ Litter causes land pollution.

4. List the **three** main parts of a plant.

    _____

    _____

    _____

5. Which is **not** a basic need of animals?

    food   shelter   water   feet

6 – 7.   Look at each column. Decide what all the words have in common. Write the best title at the top of each list. Use the choices below.

Tools   Weather   Things to Recycle

Living Things   Clouds

A)	B)
flood	cumulus
drought	cirrus
blizzard	stratus

8.   How are these animals alike?

chipmunk   bat   skunk   raccoon

A) They all live in caves.
B) They are all mammals.
C) They all walk.

Lesson #62

1. What does this symbol mean?

_____

2. Which is **not** a trait of fish?

   A) lives in water   C) lays eggs

   B) breathes with gills   D) has fur

3. An example of an amphibian is a _____.

   penguin   whale   frog   snake

4. List the **five** animal groups. One has been done for you.

   <u>mammal</u> _____

   _____

   _____

   _____

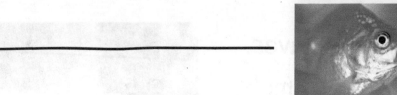

5. What is the part of the plant that grows above ground and helps hold the plant up?

A) root   C) stem

B) leaf   D) flower

6. What weather happened to make this land dry?

flood   hurricane   drought

7. Underline a natural resource. (See Lessons #49 – 50.)

         gasoline   car   water

8. Animals have **four** basic needs. List them.

_____

_____

_____

_____

Lesson #63

1. The force that pulls objects toward each other is called _____.

    natural resource   gravity   constellation

2. The word that means "to move back and forth" is _____.

    vibrate   energy   resource

3. Stars that form a pattern are called _____.

    constellations   pictures   planets

4. How are birds and fish alike?

    A) Both lay eggs.
    B) Both have feathers.
    C) Both breathe with lungs.

5. About how long does it take the moon to orbit the Earth?

    A) a month
    B) a year
    C) a day

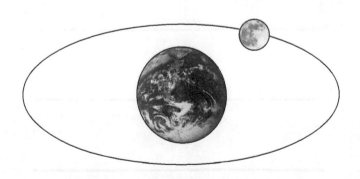

6 – 7.   Draw each part of a plant in the boxes below.

Plant Part	Drawing
root	
stem	
leaf	

8.   Which is **not** a trait of reptiles?

A) lays eggs

B) has dry skin with scales

C) gibes birth to liev young

D) breathes with lungs

Lesson #64

1. For each thing in the chart below, decide whether it is a **solid**, a **liquid**, or a **gas** and put a ✓ in the correct column.

	Solid	Liquid	Gas
tea			
oxygen			
wood			

2. Which type of cloud looks like a sheet or layer of clouds?

   cumulus   stratus   cirrus

3. Some animals go into a deep sleep for the winter. What is this called?

   A) migration
   B) life cycle
   C) hibernation

4. Which part of the plant carries water the leaves?

   roots   stem   flower

5 – 6. Look at each column. Decide what all the words have in common. Write the best title at the top of each list. Use the choices below.

Fish   Trees   Parts of a Plant

Nonliving Things   Types of Soil

A)	B)
stem	rock
leaf	straw
root	rain

7. Underline the **birds**.

cardinal   giraffe   chicken   guppy

alligator   penguin   frog   monkey

8. The plant life cycle begins with a _____.

seed   flower   leaf

Lesson #65

1. Color the part of the plant where food is made.

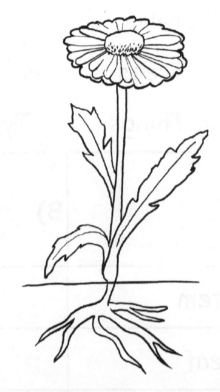

2. What are the **four** basic needs of plants?

   A) nutrients, soil, sunlight, water

   B) nutrients, water, air, sunlight

   C) sunlight, nutrients, water, shelter

3. Which type of cloud is shown?

   cirrus   cumulus   stratus

4. Something that is found in nature that people need or can use is called a(n) _____.

   natural resource   force   gravity

5. A young frog is called a _____.

Use the chart to answer the questions below.

	Breathe	Outside Covering	Babies
Reptiles	lungs	dry scales	hatched from eggs
Amphibians	gills, then lungs	wet, moist skin	hatched from eggs
Mammals	lungs	hair or fur	born alive

6. Which **two** of these animal groups lay eggs?

_____

_____

7. Which animal group has hair or fur?

_____

8. Which animal group uses both gills and lungs to breathe during its lifetime?

_____

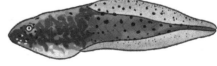

Lesson #66

Word Bank

solid	liquid	force
gravity	migration	gas
hibernation	constellation	vibrate

Across

2. a push or pull

3. fills the shape of its container

5. takes the shape of its container

6. to move back and forth

7. the trip animals make to a warmer place during the winter so they can find food

9. the deep sleep of animals during winter

Down

1. stars that form a pattern

4. has its own shape and takes up a certain amount of space

8. a force that pulls objects toward each other

Use the clues and the Word Bank to complete this puzzle.

Lesson #67

1. A _____ is a push or pull.

2 – 3. For each thing in the chart below, decide whether it is a **solid** or a **liquid** and put a ✓ in the correct column.

	Solid	Liquid
lemonade		
notebook		
glue		
can		

4. Which plant part grows underground?

   A) leaf

   B) stem

   C) root

5. The sun is a _____.

   A) planet

   B) constellation

   C) star

6. Sand, silt, and clay are types of _____.

   A) clouds
   B) weather
   C) soil
   D) plants

7 – 8. For each living thing in the chart below, decide whether it is a **mammal**, **bird**, **reptile**, **fish**, or **amphibian** and put a ✓ in the correct column.

	Bird	Amphibian	Reptile	Mammal	Fish
turtle					
penguin					
kitten					
toad					
lizard					
shark					
hawk					

Lesson #68

1. To which group does this animal belong?

    A) reptile

    B) mammal

    C) amphibian

    D) bird

2. Which tool is used to find the temperature?

    weather vane   thermometer   hand lens

3. Draw the symbol for **recycle**.

4. Plants have four basic needs. List them.

    _____

    _____

    _____

    _____

5. Which tool would you use to see something too small to see with only your eyes?

hand lens   thermometer   microscope

6. Color the part of the plant that carries water and nutrients to the rest of the plant.

7. Which of these is a **gas**?

A) paint

B) helium

C) brick

D) milk

8. Which is **not** true of an amphibian?

A) lives part of its life in water

B) lays eggs

C) makes milk for its babies

D) has moist skin

Lesson #69

1. A _____ has its own shape and takes up a certain amount of space.

liquid   solid   gas

2. Which of these is a **nonliving** thing?

sand   oak tree   grass

3 – 4. Look at each column. Decide what all the words have in common. Write the best title at the top of each list. Use the choices below.

Reptiles   Clouds   Science Tools

Living Things   Types of Soil

A)	B)
weather vane	sand
thermometer	silt
hand lens	clay

5. A _____ is a huge ball of glowing gas.

       moon   planet   star

6. Which of these is **not** something that all plants need to live?

    air   light   nutrients   soil   water

7. The sun is a _____.

   A) planet

   B) comet

   C) star

8. Name these tools.

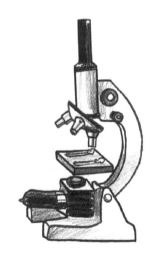

A) _____   B) _____

Lesson #70

1. Put these steps in the correct order.

    _____ Do an experiment.

    _____ Make a guess.

    _____ Observe and ask questions.

    _____ Share what you learned.

2. What are the **four** basic needs of animals?

    A) food, shelter, sunlight, water

    B) food, water, air, wings

    C) air, food, water, shelter

3. A force is a _____ or a _____.

4. Give **two** examples of a liquid.

    _____

    _____

5. What do mammals, birds, and reptiles have in common?

    A) They all have fur.

    B) They all lay eggs.

    C) They all breathe with lungs.

6. List the **five** animal groups.

_____

_____

_____

_____

_____

7. About how long does it take the moon to orbit the Earth?

    a month   a year   a day

8. Which of the following is true of amphibians?

    feathers   moist skin   hair   scales

Lesson #71

1. Draw the symbol for **recycle**.

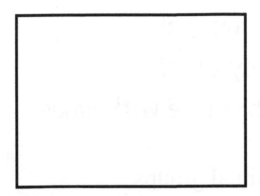

2. Look at the words below. Place each word under the correct heading in the chart.

paper   grasshopper   tulip   snow

Living	Nonliving

3. The plant life cycle begins with a _____.

   A) bean

   B) seed

   C) stem

4. Match each plant part with its definition.

    ___ stem   A) holds the plant up

    ___ root   B) where the plant makes its
                         food

    ___ leaf   C) takes in nutrients from
                         the soil

5. Which type of cloud is pictured?

    cirrus   cumulus   stratus

6. The force that pulls objects toward each other is
    called _____.

    gravity   constellation   vibration

7. A _____ can happen when there is too
    much rain.

        tornado   flood   drought

8. Which animal group lives its entire life in the water?

        reptile   bird   amphibian   fish

Lesson #72

1. Write **T** if the statement is true or **F** if it is false.

   _____ A shelter is a safe place for an animal to live.

2. What does this tool measure?

   A) temperature

   B) rainfall

   C) wind direction

3. Which type of severe weather is pictured?

   blizzard   thunderstorm   tornado

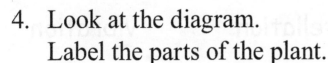

4. Look at the diagram.
   Label the parts of the plant.

   A)_____

   B)_____

   C)_____

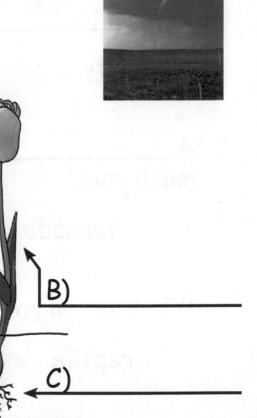

5. Underline the **mammals**.

dog   owl   turtle   squirrel

snake   penguin   toad   monkey

6 – 7. Look at each column. Decide what all the words have in common. Write the best title at the top of each list. Use the choices below.

Amphibians   Reptiles

Seasons   Weather

A)	B)
turtle	fall (autumn)
lizard	summer
alligator	spring

8. Some animals move to a warmer place during the winter so they can find food. What is this called?

life cycle   migration   hibernation

Level 2

Science

Help Pages

Help Pages
Glossary

A

air pollution — comes from factory and car fumes

amphibian — an animal group that lives part of its life in the water and part on land; lays eggs; breathes with gills when young and with lungs when an adult

B

bird — an animal group that has feathers and wings; lays eggs; breathes with lungs

C

cirrus cloud — looks like wisps of hair; highest clouds in the sky

clay — a type of soil

constellation — a group of stars that form a pattern

cumulus cloud — looks puffy; can bring strong storms

Help Pages
Glossary

drought — this happens when it has not rained for a long time and the land gets very dry and hard

fish — an animal group that lives its entire life in the water; breathes with gills; has scales on its body

flood — this happens when rivers and lakes get too high and overflow

force — a push or a pull

G

gas — a form of matter that does not have its own shape; it fills the shape of its container

gravity — a force that pulls objects toward each other

Help Pages
Glossary

H

hand lens — a tool to make something look larger

hibernation — an animal's deep sleep for the winter

L

land pollution — caused by people littering

leaf — grows out of the stem; the plant makes its food here

life cycle — all of the parts of an animal's life

liquid — a form of matter that takes the shape of its container

living thing — needs food, water, and air; it grows and changes

Help Pages
Glossary

M

mammal — an animal group with hair or fur; gives birth to live young; breathes with lungs; makes milk for its babies

matter — anything that takes up space

microscope — a tool used to see objects too small to see with only your eyes

migration — the trip (or move) an animal makes to a warmer place for the winter in order to find food

moon — a large ball of rock that travels around the Earth

N

natural resource — something that is found in nature that people need or can use

nonliving thing — does not need water, food, or air

nutrients — help a plant grow; come from the soil; food

Help Pages
Glossary

P

pollution — waste (trash) that harms the water, air, or land

R

recycle — save resources by reusing them

reptile — an animal group that has skin with dry scales; breathes with lungs; lays eggs

root — takes in water and nutrients from the soil; usually grows underground

S

sand — a type of soil

season — a time of year that has a certain type of weather

seed — the beginning of a plant's life cycle

shelter — a safe place for an animal to live

Help Pages
Glossary

silt — a type of soil

soil — made up of small pieces of rock and leftover bits of dead plants and animals

solid — a form of matter that has its own shape and takes up a definite amount of space

sound — a form of energy that you can hear

star — a huge ball of glowing gas

stem — helps to hold a plant up; carries water and nutrients to the leaves

stratus cloud — looks like a sheet or layer; lowest clouds in the sky

T

tadpole — a young frog

thermometer — measures the temperature of the air

Help Pages
Glossary

vibrate — to move back and forth

water pollution — caused by waste (trash) from factories or oil spills

weather — what the air outside is like

weather vane — measures the direction the wind is blowing

wind — moving air

Help Pages
Science Tools

We use tools to help us observe, measure, or study objects.

This tool is used to magnify an object, or make it look larger. It is called a **hand lens**.

This tool is used to measure how hot or cold something is. It is called a **thermometer**.

A **microscope** is a tool used to magnify (make larger) objects. Microscopes are helpful to see objects that are too small to see with only your eyes.

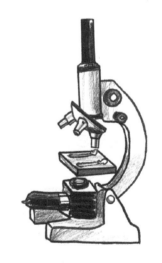

A **weather vane** is used to measure the direction that the wind is blowing.

Help Pages
Thinking Like a Scientist

When a scientist has a problem or a question, he or she uses a plan to try and find the answer. There are 5 steps to thinking like a scientist. You can use these same steps to answer your own questions about the natural world.

1. Use your five senses (sight, hearing, taste, smell, touch) to **observe** the world around you.

2. You may have a **question** about what you are observing.

3. Make a **guess** about a possible answer to your question.

4. To find the answer to your question, do an **experiment**. You can see if your guess is correct.

5. The last thing to do is to **share** what you learned from your experiment. You can write or draw pictures about it.

Help Pages
Living Things

Needs of Animals
food
water
air
shelter

Needs of Plants
nutrients
water
air
light

Life Cycle of a Plant

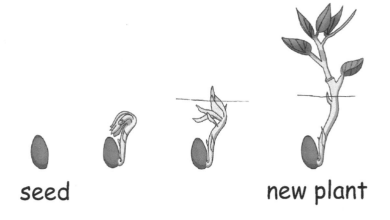

seed   new plant

Parts of a Plant

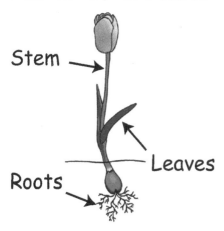

Stem

Leaves

Roots

Life Cycle of a Frog

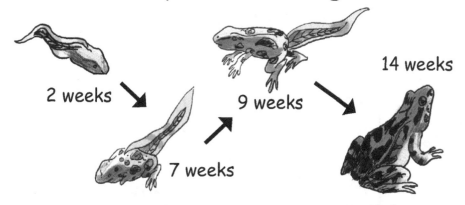

2 weeks

7 weeks

9 weeks

14 weeks

Help Pages
Animal Groups

Group	Description	Examples
Amphibians	Amphibians spend part of their life in the water and part on land. Amphibians have moist skin.	frog, toad, salamander
Birds	Birds have feathers, wings, and two legs. They breathe with their lungs and they lay eggs with a hard shell.	cardinal, robin, hummingbird, pelican, goose, duck, penguin
Fish	Fish spend their whole lives in the water. Most fish are covered with scales. Most lay eggs and breathe with gills.	trout, salmon, catfish, bluegill, bass, carp, angelfish
Mammals	Mammals have hair or fur. They use lungs to breathe. Most mammals do not lay eggs; they give birth to live young. They also produce milk for their young.	human, whale, bat, deer, dog, raccoon, rabbit, squirrel
Reptiles	Reptiles have dry, scaly skin and lay eggs on land. They breathe with their lungs. Most hatch from eggs.	snake, alligator, iguana, lizard, turtle

Help Pages
Natural Resources

Resource	Uses
water	Water is used for cooking, bathing, watering plants, swimming, and drinking.
air	People need air to breathe. Moving air can also be used as a source of energy.
plants	People use plants as food. Plants also give off oxygen. Wood from trees helps to make furniture, paper, and the houses we live in. The cotton used to make clothes comes from the cotton plant.
animals	Many people use animals for food in the form of meat, milk, and eggs. People also use animal fur for clothing.

Cloud Types

Cirrus clouds are the highest clouds in the sky. They look like wisps of hair.	
Stratus clouds are the lowest clouds in the sky. They look like a sheet or layer of clouds.	
Cumulus clouds look puffy. They can bring strong storms.	

Help Pages
Matter

Matter is anything that takes up space. Matter is all around you. The things you can see, and even the air you can't see, are matter.

	Description	Examples
solid	a form of matter that has its own shape and takes up a definite amount of space	your desk, a pencil, a banana, a car, an ice cube
liquid	a form of matter that takes the shape of its container	juice, milk, tea, shampoo, paint, water
gas	a form of matter that does not have its own shape; it fills the shape of its container	oxygen, helium